Mersenne Numbers, Strings, the Rule of 6, and the Method of Fractions

William Fidler

Bibliographic information published by the German National Library:

The German National Library lists this publication in the National Bibliography; detailed bibliographic data are available on the Internet at http://dnb.dnb.de.

ISBN: 9783346936646
This book is also available as an ebook.

Print and binding: Books on Demand GmbH, Norderstedt, Germany
Printed on acid-free paper from responsible sources.

The present work has been carefully prepared. Nevertheless, authors and publishers do not incur liability for the correctness of information, notes, links and advice as well as any printing errors.

GRIN web shop: https://www.grin.com/document/1389654

Mersenne numbers, Strings, the Rule of 6, and the Method of Fractions

W M Fidler

Abstract

The work developed here brings, amongst other things, a significant measure of order to the search for prime numbers.

It was shown in [1] that the prime numbers could be confined to one row of a matrix and, from the construction of strings of consecutive numbers called primary strings from this matrix, it was determined that prime numbers could only be found throughout the range of the integers on either side of any number which was divisible by **6**. Indeed, it was considered appropriate to add a statement to this effect to the definition of a prime number. A procedure is devised that we have called, the Rule of **6,** whereby the location of a number within the string can be determined by examining the numerator of the fraction constituting the remainder of any number therein when divided by **6**. These numerators we have termed, the characteristic signature of a number, and are found to be of great utility in the search for a prime number.

It is shown conclusively that no Mersenne number can have an integer square root, this being one of the current unsolved problems in Number theory. Further, it is found that only one Mersenne number may be found within any primary string and, if it has the characteristic signature of **3/6**, can never be prime. Only Mersenne numbers having the characteristic signature of **1/6** can be prime but, as in the case of the pseudo prime number generating function, this does not guarantee that the number is prime. All Mersenne numbers are shown to have only either of two characteristic signatures, viz. **1/6** and **3/6.**

Within the limited range of the computational facilities available to the author it is found that we may generate a set of numbers from the formula, $F = (M + 2)/3$, where M is a Mersenne number. It is found that whenever **M** is prime then **F** is prime. However, primes may sometimes be generated from this expression when the index of the Mersenne number is prime but the number itself is composite.

An accelerated version of trial division using a process that we have called the Method of Fractions is developed, and from which we can construct, what are essentially one row matrices and which we have called templates. Each of these is associated with a parameter called the base number which may be factor of a number of interest; the range of a template, which may be extended by a simple process, is equal to the number of digits of any number of interest; moreover, increasing the extent of a template by one digit increases the range of the numbers which may be examined by an order of magnitude.

List of Contents

Introduction

Until the latter half of the 20th century prime numbers were considered to be the domain of the number theorist and whilst of great philosophical interest were considered to have little, or no relevance outside of pure mathematics. Indeed, G H Hardy prided himself upon the fact that his work could have no relevance in the wider world, particularly for applications associated with the war of 1914 – 1918.

In the nineteen seventies it was found that the properties of very large numbers, and prime numbers in particular, could be employed in the encryption of sensitive financial and other data. This led to an explosion of interest in these numbers, culminating in the Great Internet Mersenne Prime Search (GIMPS) [2] which continues to the present time. It is now a badge of distinction to determine the largest prime number. Much attention is concentrated upon the Mersenne numbers for they can be determined by a simple formula and grow exponentially. Numbers of this kind have been investigated for centuries before Mersenne's time and the literature associated with prime numbers from them literally teems with prime numbers of different types (although, at bottom, they are all primes).

The work presented here, whilst focusing upon Mersenne numbers is not limited to these and has universal application.

Analysis

It was shown conclusively in [1] that the prime numbers can only exist at the positions occupied by the second and penultimate numbers in the strings developed from the jocularly-titled, Magic Matrix. Given the nature of the strings it then follows that prime numbers can only exist on either side of a number which is divisible by six.

We now present three sets of numbers from [1] where the prime numbers are presented in bold and which perform the dual function of providing the reader with sufficient data to explore some of the findings of the work, whilst demonstrating the major simplification which results from the forming of the Magic Matrix. Simply for reasons of space the tables are only of utility for relatively small Mersenne numbers.

3	**5**	**7**	9	**11**	**13**	15	**17**	**19**	21	**23**	25	27	**29**	**31**
33	35	**37**	39	**41**	**43**	45	**47**	49	51	**53**	55	57	**59**	**61**
63	65	**67**	69	**71**	**73**	75	77	**79**	81	**83**	85	87	**89**	91
93	95	**97**	99	**101**	**103**	105	**107**	**109**	111	**113**	115	117	119	121
123	125	**127**	129	**131**	133	135	**137**	**139**	141	143	145	147	**149**	**151**
153	155	**157**	159	161	**163**	165	**167**	169	171	**173**	175	177	**179**	**181**
183	185	187	189	**191**	**193**	195	**197**	**199**	201	203	205	207	209	**211**
213	215	217	219	221	**223**	225	**227**	**229**	231	**233**	235	237	**239**	**241**
243	245	247	249	**251**	253	255	**257**	259	261	**263**	265	267	**269**	**271**
273	275	**277**	279	**281**	**283**	285	287	289	291	**293**	295	297	299	301
1	2	3	4	5	6	7	8	9	10	11	12	13	14	15

2013 2015 **2017** 2019 2021 2023 2025 **2027** **2029** 2031 2033 2035 2037 **2039** 2041

2043 2045 2047 2049 2051 **2053** 2055 2057 2059 2061 **2063** 2065 2067 **2069** 2071

2073 2075 2077 2079 **2081** **2083** 2085 **2087** **2089** 2091 2093 2095 2097 **2099** 2101

2103 2105 2107 2109 **2111** **2113** 2115 2117 2119 2121 2123 2125 2127 **2129** **2131**

2133 2135 **2137** 2139 **2141** **2143** 2145 2147 2149 2151 **2153** 2155 2157 2159 **2161**

2163 2165 2167 2169 2171 2173 2175 2177 **2179** 2181 2183 2185 2187 2189 2191

2193 2195 2197 2199 2201 **2203** 2205 **2207** 2209 2211 **2213** 2215 2217 2219 **2221**

2223 2225 2227 2229 2231 2233 2235 **2237** **2239** 2241 **2243** 2245 2247 2249 **2251**

2253 2255 2257 2259 2261 2263 2265 **2267** **2269** 2271 **2273** 2275 2277 2279 **2281**

The Magic Matrix

3	4	**5**	6
9	8	**7**	6
9	10	**11**	12
15	14	**13**	12
15	16	**17**	18
21	20	**19**	18
21	22	**23**	24
27	26	25	24
27	28	**29**	30
33	32	**31**	30
33	34	35	36
39	38	**37**	36
39	40	**41**	42
45	44	**43**	42
45	46	**47**	48
51	50	49	48
51	52	**53**	54

It is conjectured that extending the matrix indefinitely will show that the prime numbers are embedded in the third row of the matrix and can only be located on either side of a number which is divisible by **6**. Whilst this verifies the prime number generating function, given by the formula, $p = 6\,q \pm 1$, $q = 1, 2, 3, etc$ it also reveals that the function is not infallible in the prediction of the prime numbers and, in the opinion of the author should more correctly be described as the pseudo prime number generating function. Nevertheless, what we have established is that we need only investigate numbers in the locations described, for all of the other numbers cannot be prime.

Primary strings and the Rule of 6

As in [1] it was shown that sequences of consecutive numbers could be extracted from the Magic Matrix and which were named primary strings. An example of this is the initial sequence **6, 7, 8, 9, 10, 11, 12**. It may be observed that there are two prime numbers in this sequence, **7** and **11**, although it is not guaranteed that the numbers at these locations in other primary strings are prime numbers. However we note that if we divide all of the 'interior' numbers within the string by **6**, they have remainders of, **1/6, 2/6, 3/6, 4/6, 5/6.** We term this procedure the Rule of **6.** These fractions, which we call characteristic signatures, are of great significance for they allow us to locate any number within its respective primary string simply by examining the fractional remainder resulting from division of the number by **6.** Moreover, in the search for prime numbers we can then eliminate all numbers which do not have either **1/6** or **5/6** as a remainder.

Division of a number by, for example, **6** is an easy process but, as the numbers become larger the propensity for error in longhand division increases and reversion to an electronic calculator is then a logical next step. However, as the number under examination becomes ever more large, the range of the calculator display is insufficient to yield a remainder and so the characteristic signature, and hence the location of the number in a primary string cannot be determined. We now proceed to devise a process which will permit the determination of the fractional remainder for any number.

The Method of Fractions

Rather than proceed to develop this concept expressed in arcane mathematical format we adopt an approach, characteristic of much of previous work by the author, and explain the process by example.

Consider the decadal sequence **10^9, 10^8, 10^7, 10^6, 10^5, 10^4, 10^3, 10^2, 10^1, 1.**

Let us divide each of the terms by **6** starting at the RH end of the sequence. The numerators of the remainders of this procedure will then be, **1, 4, 4, 4, 4, 4, 4, 4, 4, 4.** We call the dividing number, in this case, **6,** the base number. We now choose a number, at random and show how to determine its location in a primary string.

Consider the number **1253964723.** Starting from the left we now write all of the above numerators in the following manner: **4 4 4 4 4 4 4 4 1** and immediately beneath it the random number,   **1 2 5 3 9 6 4 7 2 3.**
We now form the undernoted sums.

$1 \times 4 = 4$, $2 \times 4 = 8$, $5 \times 4 = 20$, $3 \times 4 = 12$, $9 \times 4 = 36$, $6 \times 4 = 24$, $4 \times 4 = 16$, $7 \times 4 = 28$, $2 \times 4 = 8$, $1 \times 3 = 3$. We now sum the results on the RHS of these simple multiplications and divide by **6,** i.e.
$139/6 = 23 + 1/6.$**

It is then concluded that this number is located in the second position of a primary string and may be prime. We can verify that this location is correct, for, if we subtract **1** from the sum of the multiplications we get **$138/6 = 23$,** i.e. there is no remainder and this is the characteristic signature of the numbers which bound a primary string.

Indeed, in this case of division by **6** it is not necessary to carry out the individual multiplications for all that is required is that, starting from the left hand side we sum the digits of the number as far as the penultimate number, multiply this sum by **4,** add the last digit and then divide by **6.**

In Number theory there is much interest in very large prime numbers. It is then entirely possible that the magnitude of the **sums** that result from the procedure described above could exceed the display of an electronic calculator. This is easily resolved by simply applying the whole process to this sum and then dividing that result by the base number. Mechanisation of the procedure is left to practitioners of computer coding.

We now present a table for a selection of base numbers, denoted by, **k**, and which may be applied to any number, N in the range $6 \leq N$ $(10^{12} - 1)$, (although the reader can choose to extend the range to any desired upper limit, **x**).

	10^x	r	r	r	r
	10^12	1	0	4	1
	10^11	1	0	4	5
	10^10	1	0	4	4
	10^9	1	0·	4	6
	10^8	1	0	4	2
	10^7	1	0	4	3
	10^6	1	0	4	1
	10^5	1	0	4	5
	10^4	1	0	4	4
	10^3	1	0	4	6
	10^2	1	0	4	2
	10	1	0	4	3
	1	1	1	1	1
k		3	5	6	7

To illustrate the use of the table let us devise a number which we know has a factor; we choose the number **286489** (equal to **7 x 40927**) and ask if **7** is a factor of that number.

We then form the following quantities and their sum:

9x1 = 9, 8x3 = 24, 4x2 = 8, 6x6 = 36, 4x8 = 32, 2x5 = 10. We now sum the results of these multiplications and divide by **7**. Hence, **119/7 = 17.** There is no remainder and therefore **7** is a factor of **286498.** If our purpose is to search for prime numbers then we have no interest in any other factors**.**

The reader is invited to extend the table to any size and may thus investigate numbers for primality (or otherwise).

It is of interest to note that the above table gives an explanation of why, if the sum of the digits of a number is divisible by, **3,** then that number cannot be prime. Further it is also obvious why, any number ending in **5** cannot be prime.

Attention is also drawn to the repeating pattern of numbers in the last column, although such patterns cannot be guaranteed for future columns. However, we emphasise the utility of the this tabular approach by showing that the regularity in some of the sequences presented in the above table may not be solely confined to the lower end of the number spectrum. Without constructing another table, consider the sequence of the **r** resulting from division by **11.** It is easy to see that this sequence, starting from the lower end is:

1, 10, 1, 10, 1, 10, 1, 10, 1, 10, 1, 10, 1, 10, 1, 10, 1, 10, 1, 10. Purely for the purpose of illustration, the sequence has been terminated at the **20**th member. We now have a simple procedure for determining whether **11** is a factor of any **20** digit number. Consider the following **20** digit number chosen at random.

3 6 8 7 3 5 9 6 2 1 4 5 2 1 8 7 6 5 17. We now write this in reverse order in the form shown:

$$7\ 5\ 7\ 1\ 5\ 1\ 6\ 5\ 7\ 6$$

$$1\ 6\ 8\ 2\ 4\ 2\ 9\ 3\ 8\ 3.$$

The numbers in the top row are those that would be multiplied by **1,** whilst those in the bottom would be multiplied by **10.** We simply sum the numbers in the top and bottom rows respectively to give **50 x 1** and **46 x 10,** and then add the two sums and divide by **11** to give **(50 + 460)/11 = 510/11 = 46 4/11.** The result has a non-zero remainder, and hence **11** is not a factor of **36873596214521876517.**

As noted above, the extent of this table may be expanded without limit. Determination of the multipliers therein is a simple process for all that is required is that new numbers may be generated proceeding in the upward direction by taking the preceding number, multiplying it by **10,** and then dividing by the base number and retaining the numerator of the fraction.

It is somewhat ironic that we can determine whether a base number is a factor of another number without dividing the base number into that number..

The above example reveals that we may determine any **20** digit number which is divisible by **11.** We proceed as follows:

Consider the number **510** in the above example; the nearest number greater than this which is divisible by**11** is **517.** This can be obtained from the sum: **47 x 1 + 47 x 10 = 517.**

We see that this may be done by increasing the sum of the lower row by **10** whilst decreasing the sum of the upper row by **3.** Now, we could make these changes anywhere in the rows, but since we wish to obtain the nearest number above the **20** digit number which is divisible by **11,** then we make these changes at the RH end of the number. Hence, the nearest number which is greater than **36873596214521876517** and is divisible by **11** is **3687359621452187652<u>4</u>.** By a similar process we can determine the nearest lower number which is divisible by **11,** for all that is required is that the sum of the sums of the two rows is **506 (i.e. 517 – 11).** This is obtained by subtracting **4** from the sum of the first row. Hence, the nearest number lower than the **20** digit number which is divisible by **11** is **3687359621452187651<u>3</u>.**

Templates

We now define the columns headed by **r** in the previous tables as templates, and, if we imagine each template to be a solid object then, if this object is rotated about its base by $\pi/2$ in the positive sense and the number of interest aligned underneath, then we have a **2 x z** matrix where **z** is the number of digits in the number of interest. Of course, it is implied that the extent of the template and the number of digits in the number of interest are the same.

We now form products of corresponding elements of the rows and sum these products. If division of this sum by the base number does not yield an integer, then the base number is not a factor of the number of interest. If typographical space is not an issue then we may retain the template in its original position and set alongside it the number of interest written in ascending order, and then perform the same operations as noted above. In this instance there then results a **z x 2** matrix.

The template is so-called for, in a sense, it is fitted onto the number of interest.

It is considered that we have found a highly-accelerated and powerful form of trial division for, if we use the previous demonstration as an example, then, for a 20 digit number we may examine that number to see if the base number is a factor simply by forming the relevant template. Any changes that we may make to the number do not require repetition of the long (and tedious) process associated with long division for all that is required is that we produce the new sums of the products and divide the total sum by the base number.

Mersenne numbers

These numbers, which we denote by, **M,** are generated by the simple formula, $M = 2^n - 1$ and have been studied from antiquity. Just like the pseudo prime number generating function, $6q \pm 1$, the Mersenne formula can frequently, but not invariably, generate prime numbers.

We now investigate Mersenne numbers using the concepts of primary strings, the Rule of **6** and the Method of Fractions.

It was shown in [1] that any odd number, **N,** could only have an integer square root if it satisfied the equation, $(N - 1)/4 = ab$, where **a** and **b** are consecutive smaller numbers; if the equality is satisfied then the square root of **N** is equal to the sum of **a** and **b.**

It is stated in [3] that all known Mersenne prime numbers with **n** prime are square-free and hence it is conjectured that all Mersenne primes are square free. However, Guy, in [4] expresses his belief that this is not the case and that there exists some Mersenne primes which have integer square roots. Given that this is considered to be one of the unsolved problems of Number theory, consider the following argument:

Let $N = M = 2^n - 1$, therefore, we may write; $2^{n-1} - 1 = 2ab$. Now, 2^{n-1} is always an even number, and so, $2^{n-1} - 1$ must be odd. The quantity **2ab** is always an even number, hence, we conclude that no Mersenne number, can ever have an integer square root; indeed we could express this in the following theorem: '**A function of the form $N = 2^n - 1$, where n and N are integers, prime or otherwise, can never have an integer square root.**'

We showed earlier that the location of any number within a primary string could be determined by division of that number by **6.**

Let us write $M/_6 = 2^n/_6 - 1/_6$. We now construct the following table, using the Method of Fractions and denote the numerator of the fractional remainder of $2^n/_6$ by the symbol, **r.**

n	1	2	3	4	5	6	7	8	9	10	11	12	13
2^n	2	4	8	16	32	64	128	256	512	1024	2048	4096	8192
r	2	4	2	4	2	4	2	4	2	4	2	4	2
-1	1	3	1	3	1	3	1	3	1	3	1	3	1

The values in the last row are obtained by subtracting 1 from the numbers in the row above which denote the numerators of the characteristic signatures of the contents of a string. Hence, we see that for n odd the Mersenne number will be located at the second position in a primary string, whereas, for n even the Mersenne number will be located at the central positon and, moreover, that number can never be a prime number. We then have shown that, in the context of primary strings, Mersenne numbers can only be found in two distinct places

12

throughout the whole range of the counting numbers, and have characteristic signatures of **1/6** and **3/6**.

There is no reason to expect that the pattern **1 3 1 3 1 3** will ever change, for successive **r** are determined by multiplying the preceding **r** by **2** and dividing the result by the base number and retaining the numerator of the resulting fraction.

There can never be more than one Mersenne number in a string, for successive Mersenne numbers are separated by a factor of 2^n where **n** is the index associated with the lower Mersenne number, and, in the context of a primary string, 2^n always exceeds the dimension of the string. Indeed, it follows that the separation between Mersenne numbers becomes progressively larger as **n** increases.

We now equate the pseudo prime number generating function to the Mersenne number.

Hence, we have, $6q + 1 = 2^n - 1$, where **q = 1, 2, 3**, etc.

Rearranging this gives, $q = {}^{2^n}/_6 - {}^2/_6$. We can use the previous table to identify values of the index **n** which are consistent with integer values of **q**.

This is accomplished by simply subtracting **2**, which is the numerator of the second term on the RHS of the above expression.

n	1	2	3	4	5	6	7	8	9	10	11	12	13
2^n	2	4	8	16	32	64	128	256	512	1024	2048	4096	8192
r	2	4	2	4	2	4	2	4	2	4	2	4	2
-2	0	2	0	2	0	2	0	2	0	2	0	2	0

We see that integer values of **q** only exist at odd values of **n**.

We stated previously that, just as in the case of the pseudo prime number generating function, there is no guarantee that the Mersenne formula will invariably produce a prime number. To illustrate this we now apply the Rule of **6** to a selection of Mersenne numbers which are known to be composite.

n	$2^n - 1$	R
23	8388607	23 + 1/6
29	536870911	26 + 1/6
37	137438593471	36 + 1/6
41	2199023255551	32 + 1/6

Here, **R** is the result of applying the Rule of 6..

It will be seen that, since the fractional part of **R** is **1/6**, then the Mersenne numbers are located at the second position in their respective primary string.

Hence, although the numbers occupy one of the only two positions throughout the whole range of the counting numbers at which a number may be prime, they are composite.

We have shown [1] that if we add the numbers in the second and penultimate positions in a string and divide the result by **6** we get an odd number, and that a string, but not every string can only contain one Mersenne number, either at the mid position, where it can never be prime or at the second position, where it might be prime. Hence, we now examine strings where the number occupying the second position is a Mersenne number, **M**.

Forming the sum mentioned above, we have $[M + (M + 4)]/6 = N = (M + 2)/3.$

Since we are dealing with an expression associated with primary strings the lowest Mersenne number that we may investigate is $M_3 = 7$. Now this is a Mersenne prime and so, applying the above expression to **7** we get, **9/3 = 3**, which is prime. Now, this could be simply fortuitous and so, consider the next Mersenne prime which is $M_5 = 31$. We see that the above formula then yields **33/3 = 11**, which again is prime.

If we consider $M_8 = 255$, which is obviously not prime for the sum of its digits is divisible by **3**, we obtain **257/3**, which is not even an integer.

We now apply the above test to Mersenne numbers which are known to be prime.

$M_7 = 127 \rightarrow \frac{127+2}{3} = 43$, which is prime.

$M_{13} = 8191 \rightarrow \frac{8191+2}{3} = 2731$, which is prime.

$M_{17} = 131071 \rightarrow \frac{131071+2}{3} = 43691$, which is prime.

$M_{19} = 524287 \rightarrow \frac{524287+2}{3} = 174763$, which is prime.

$M_{31} = 2147483647 \rightarrow \frac{2147483647+2}{3} = 715827883$, which is prime.

The Mersenne prime derived by Lucas in **1876** and which we denote as M_L remained the largest known prime for almost a century and is given by the formula, $M_L = 2^{127} - 1$.

Hence, $M_L = $ **170141183460469231731687303715884105727,**

Now, $(M_L + 2)/3 = $ **56713727820156410577229101238628035243.**

If this number is divided by **6**, using the template with **6** as base number the result has a remainder of **1/6**, thus showing that this number has the characteristic signature of a number occupying the second position in a primary string, which is a possible location for a prime number. Lack of computational facilities prevents the determination of the primality of this number.

The reader is invited to extend this approach to all of the known Mersenne primes. It is conjectured that this will produce a wholly new set of prime numbers; upon division of these by **6** it will be seen from their characteristic signatures that they lie immediately on either side of a number which is divisible by **6.**

Being utterly without humility the author names the numbers generated in this way, Fidler numbers, denoted by **F.**

Hence we have, $F = (M + 2)/3.$

Over the limited range of Mersenne primes that has been examined, if **M** is prime then **F** is prime. However, let us consider $M_{11} = 2047;$ this is the smallest Mersenne number which is composite and has a prime exponent $(n = 11).$

If we apply the above formula with $M = M_{11}$ then we get $F = 683,$ which is prime. This should not be taken to indicate that all Mersenne numbers with prime exponent and which are composite can be substituted into the above equation and obtain another prime, for whilst M_{23} will yield another prime i.e. **2796203,** substitution of M_{29} results in the composite number, **178956971;** however, it should be noted that both Mersenne numbers are composite. Indeed, it is found, when examining the Mersenne numbers over a limited range of the index, **n,** that when **n** is composite the formula always yields either a non-integer or a composite number. It is conjectured that this is a universal feature, but its verification, or otherwise is considered a subject for further study.

All of the immediately above is a timely reminder that when dealing with prime numbers, nothing should be taken for granted.

Discussion

It is considered that the work presented here brings a significant measure of order to the investigation of the disposition of prime numbers within the range of the counting numbers.

The confining of all of the prime numbers to the third row of the Magic Matrix, represents a major simplification in itself when the seemingly random distribution of the primes in the data shown at the opening of the work is contemplated. The Rule of **6,** amongst other things, permits the identification of the location of any number within a primary string, the nature of that number, and is central to showing that Mersenne numbers can only be found in two places within a primary string throughout the whole range of the integers. These Mersenne numbers have characteristic signatures of **1/6** and **3/6,** respectively; the latter signature is that of Mersenne numbers which can never be prime. Moreover, it is shown that a primary string can never contain two Mersenne numbers.

The conjecture that Mersenne numbers can never have integer square roots has, until this time, to the best of the author's knowledge, been considered one of the unsolved problems of Number theory. We now regard this question as settled.

The accelerated version of trial division which we have called for brevity, the Method of Fractions is really a Method of Remainders expressed as Fractions, and this, together with the concept of templates and their application is, to the best of the author's knowledge, novel. The templates, which are peculiar to a base number may be of any extent, and once formed could be placed in a library of templates in the memory of a computer. Moreover any template may be extended by the simple rule noted in the work, the extent of the template being changed to accommodate the magnitude of the number being examined for factors. It may be noted that increasing the extent of a template by one digit increases, by an order of magnitude the numbers that may be examined. If the object of the exercise is to search for prime numbers, then if any factor is found when proceeding in a systematic fashion of increasing base numbers, then we have no interest in any other factors. The investigation of the sequences of numbers of which the templates are comprised is a complete enterprise in its own right ---but this is an endeavour postponed to another day.

The conjecture that the numbers that we have called the Fidler numbers are sometimes prime, but are always prime when obtained from the formula $= (M_p + 2)/3$, where M_p is a Mersenne prime has, for the lack of suitable computing facilities, not been pursued and further investigation is left to those with the computational facilities so to do.

W M Fidler

August 2023.

16

References

[1] A new perspective on the determination of the prime numbers.

      W M Fidler

      GRIN Verlag Cat No. v1360641

      May 2023

[2] Great Internet Prime Search (GIMPS).

      Wikipedia

[3] Wolfram Mathworld

[4] Mersenne Primes, Repunits, Fermat numbers, Primes of the Shape $k\,2^n + 2$.

      R K Guy

      Unsolved problems in Number Theory, 2nd Ed. A3, pp 8-13.

      New York, Springer Verlag, 1994.